Plotting and Building
the Pyramids of Giza

Daniel Gerardo

2018

ISBN-10:1483993876
ISBN-13: 978-1483993874

Published 6/3/2018

To build the Great Pyramid, it must first be plotted.

Daniel Gerardo

DANIEL GERARDO

CONTENTS

INTRODUCTION:

Monuments of human strength and ingenuity, the pyramids have always compelled admiration and aroused curiosity. Erected on the Giza Plateau, Pharaoh Khufu's pyramid (Cheops in the Greek designation) or Great Pyramid, as it is called, is a masterpiece of construction. Together with the pyramids of the pharaohs Khafre and Menkaure, the Sphinx, mastabas and satellite pyramids, it is part of the funerary complex of Giza.

This work – regarded as the first wonder of the ancient world – is the result of the evolution in construction occurring in the pharaonic tombs of the Ancient Egyptian Empire.

Those who constructed the pyramids learnt to build them in Egypt based on their own experience. As the Egyptian civilization built the pyramids, the pyramids built the Egyptian civilization. The discovery of the builders' village in Giza by Zahi Hawass and Mark Lehner as well as the Diary of Merer – foreman of the Great Pyramid works – , which describes how the casing stones were transported, helps us to place the work in the historical context .

In this evolution of construction, two requirements stand out that were of significance to the pharaohs:

a. The requirement of shape: The perfection in the plotting of the shape of the smooth-faced pyramids begins with the pyramids built by Pharaoh Sneferu and reaches levels of excellence in the pyramids of Giza.
b. The requirement of height: The height of the pyramids built gradually increased. In the great pyramids of Giza there is a large increase in height which is a feat of ancient engineering.

The surveyors of ancient Egypt worked out the plotting of the Great Pyramid with such precision and accuracy that it was only possible to reproduce them with the use of modern surveying instruments. The existing proposals do not help to understand or reproduce the

1

astonishing accuracy achieved in the design of the great pyramids of Giza.

The aim of this book is to present my conclusions after a decade's worth of research on the subject.

There are a wide range of theories as to the building of these pyramids, but they deal exclusively with the requirement of height. These theories basically analyze how the blocks were transported and raised to a great height during the construction of the pyramids. The survey of the pyramid is mentioned simply as a side issue and with no solution, due to the lack of historical documents, and with no significant impact on the building erected. However, the construction requirement of shape is not merely a technical detail. Like all civil works, the construction of the great pyramids of Giza required working out how to survey them in the first place.

To build the Great Pyramid, it must first be plotted.

To think that it is possible to build the pyramid without plotting it first is a misconception of our times that the ancient Egyptians would not understand and the Pharaoh would not authorize.

As we shall analyze later, plotting the pyramidal shape conditions the construction stages of the pyramid. This is the reason why in order to build the pyramid you must first plot it with the required accuracy, and we cannot skip this construction requirement. This point is key to understanding the construction. Trying to build the pyramid without plotting it has led to a maze of theories that basically address the requirement of building the highest pyramid. The only way to overcome this confusion, the labyrinth of theories and scepticism and be able to understand the construction of the pyramids is by just starting at the beginning and plotting the pyramid.

But, how do you plot the Great Pyramid with that amazing accuracy without measuring with accurate instruments? How do you plot the base of the pyramid with 230-metre sides with an average error of 15 mm in length and 32 seconds in the angles? How do you achieve such high precision in measurement without using optical instruments?

In addition, logic and the very foundations of metrology (the scientific study of measurement) say that the greater the distances to be measured, the greater the errors. However, the opposite is true of the Egyptian pyramids. The largest pyramids are the most accurate.

Another amazing fact is that there is no large building in Egypt that has the accuracy of the great pyramids. It is quite obvious that if they had developed instruments to help them measure with such precision and accuracy, they would have used them on other buildings as well. These amazing and seemingly illogical facts are a sign that the pyramids were not plotted by measuring, and they give us certain insights into how they did it. For instance, we know that with the method they used, the larger the pyramid, the greater the accuracy achieved. Furthermore, it is a technique that was only applied to the plotting of the pyramids, since it was not used with other buildings.

We also know that the accuracy in plotting starts with the smooth-faced pyramids, and we can distinguish two types of surveying:

a. Inaccurate surveying, used on the stepped pyramids, consistent with the measuring tools existing at the time.
b. Accurate surveying, developed through the plotting of smooth-faced pyramids.

Accurate plotting began to develop with the smooth-faced pyramids, and the clearest clue as to the technique used for this plotting is the deviation with which they plotted the great pyramids in relation to the cardinal points.

This deviation of the bases of the pyramids has traditionally been regarded as an orientation error, but it is not. It is simply the result of applying the technique that they used, which we shall describe in this book. The application of the technique used will help us to plot the Great Pyramids of Giza with the required accuracy and reproduce the characteristics of the original plotting.

We will then be in a position to answer other old questions:

What was the original height of the Great Pyramid? What was the original relation between the height and the base of the pyramid?

Finally, only after identifying the construction stages of the building, which are the results of the plotting used, will we be able to analyze the techniques used to raise the blocks in each of them.

Understanding how the pyramids were plotted will enable us to visualize without much difficulty how they were built, with the simplicity and efficiency that characterized the ancient Egyptians.

The Author

CHAPTER I
Evolution of Pyramids

In the course of human history, various civilizations have built pyramids for religious and funerary reasons. The shared purpose of these civilizations was to erect tall buildings. Building with stone blocks, they came up with similar architectural solutions.

If one starts from a square base and the goal is to build something tall using stone blocks, the only stable structure possible is a pyramid. It was necessary for man to develop materials such as steel and cement to build at great heights with different shapes.

The Great Pyramid is the tallest and best built pyramid; the one for which the construction requirements were met with extraordinary accuracy. This masterpiece is the result of an evolution in construction that began with the mastabas and reached its peak in this pyramid. We can assert without a shadow of a doubt that those who constructed the Great Pyramid learnt to build it in Egypt because it is clearly documented in the pyramids built. As they often say, the Egyptian civilization built the pyramids ... as the pyramids built the Egyptian civilization.

Five milestones in construction are identified in this evolution that are signs of the progress made and enabled the construction of this masterpiece.

• First milestone Mastaba

• Second milestone Stepped Pyramid

• Third milestone Smooth-Sided Pyramid

• Fourth milestone The Great Pyramid

Figure 1: Evolution of the Pyramids

The Stepped Pyramid

Pharaoh Djoser's tomb is the first pyramid built during the Old Kingdom of Egypt and marks the beginning in the evolution of royal tombs, from the mastabas to the great pyramids.

The mastabas were built during the first kingdoms with adobe bricks, the use of stone being restricted to isolated sectors of the buildings (see Fig. 2). The stepped pyramid was entirely made of stone, and it is the building for which limestone was first used on a large scale (see Fig. 3). The construction of this pyramid is ascribed to the sage Imhotep, Pharaoh Djoser's vizier, regarded by Manetho as the inventor of the art of building with stone (Lehner 1997: 84).

Figure 2: El-Faraun Mastaba

The height of the resulting pyramid is 60 m, while the base measures 121 x 109 m. (Edwards 993: 37).

Surveying the Stepped Pyramid

a. Orientation of the Pyramid and Plotting of the Reference Line:

The survey of the stepped pyramids is consistent with the inaccurate tools that they used to plot them. The plotting started with the drawing of the "reference line," because this line will provide the square for the base with its axes and, in turn, the rest of the structure.

The reference line is drawn in the direction of the north-south or east-west cardinal points. The method most widely accepted by specialists to obtain the reference line involves determining true north by means of the rising and setting sun, or another star, in relation to the centre of a circle from which it is observed. The bisection of the arc formed in the circle indicates true north (Edwards 1993: 251).

Figure 3: Stepped Pyramid of Djoser

Another suggested procedure involves marking the shortest shadow cast by a gnomon (pole) in the course of a day. The word gnomon comes from the Greek and means 'indicator.' It designates a vertical shaft casting a shadow. This is the so-called solar noon shadow, which occurs when the sun has reached its highest position on that day in the cardinal point south (azimuth 180°), casting the shadow of the pole due north. The result of joining the dots of the shortest shadow obtained over the course of several days as the shadow moves is a North-South line. (Smith 2006: 80).

Kate Spence proposes that the ancient Egyptians used the stars, a method known as "simultaneous transit," to align the pyramids with due north. She further argues that the deviation from the cardinal points in the orientation of the pyramids is due to the phenomenon known as Earth's precession, which helps to determine the date on which they were made. In Glen Dash's opinion, Spence's simultaneous transit theory was brilliantly conceived. However, the available evidence, when viewed collectively, does not support it. (1).

Martin Isler proposes that the Egyptians used the sun for the orientation of the pyramids. He believes that they used a technique known as the "Indian Circle Method". (2) Glen Dash made the practical demonstration of this method, and he obtained a deviation in the alignment that was similar to – and even smaller than – that of the ancient Egyptian surveyors.

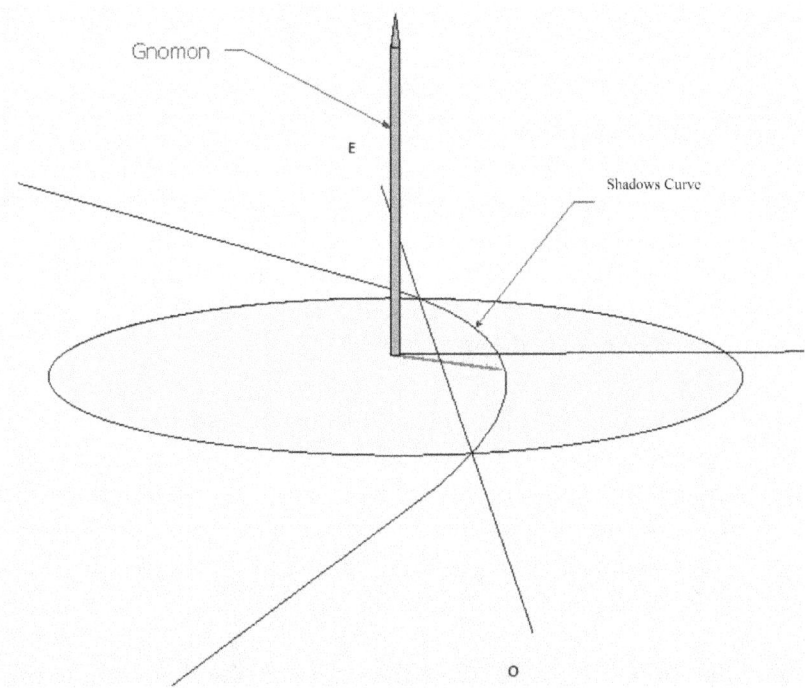

Figure 4: The Indian Circle

The Indian circle method involves placing a vertical gnomon on a levelled surface on which the curved path of the shadow cast by the gnomon during the summer solstice will be drawn. Next, a circle is drawn with its centre in the gnomon and its radius equal to its height.

The circle will intersect the curved shadow path at two points, A and B. Next, draw a line joining these points to get the reference line. This line will be orientated in accordance with the path of the sun within the interval when the path was drawn. If the path is drawn on or near the summer solstice day, the sun will have no declination, and the reference line will have an East-West orientation. If it is drawn any other day away from the solstices, the path of the sun will have a declination and the reference line will have a deviation from the cardinal points that will be related to the sun´s declination. If that day of the year the sun is moving away from the summer solstice, the deviation of the reference line will be counter-clockwise, while if it is moving closer, it will be clockwise.

This technique has a logical basis that we shall work out later during the plotting of the Great Pyramid.

b) Plotting the square for the base:

 The reference line that has been obtained is used for drawing the square for the base and its respective axes on the levelled surface, for which it is necessary to draw perpendicular lines accurately. The most widely accepted method for drawing perpendicular lines involves using the 3-4-5 triangle (Lehner 1997: 213).

A string measuring 12 units in length is fastened to a stake located at point a. By measuring 3 units on the baseline you get point c. Point d is determined by extending the string 5 units; its location is perfectly determined by placing the end of the string on point a. By joining point a with point d, you get the a-d line, perpendicular to the a-b baseline. (see Fig.: 5)

Using the same procedure, draw the opposite line that completes the north side of the base. The lengths are measured with a rod whose length is established in royal cubits.

Once the square base has been drawn, we are in a position to build the first step and the rest of them, with their corresponding treads, all the way to the top.

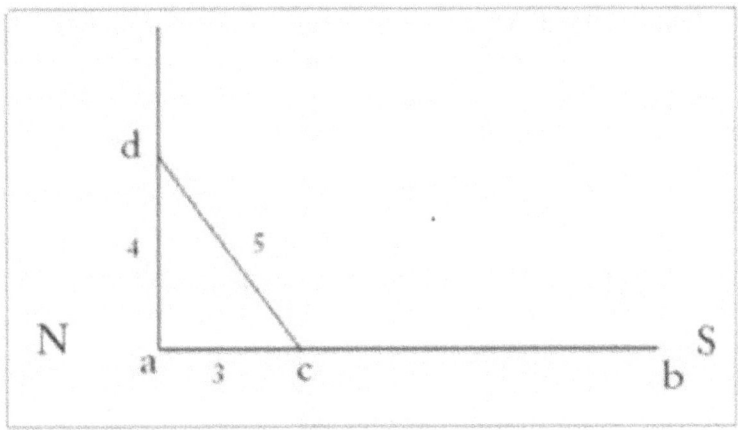

Figure 5: Drawing a right angle

The Smooth-Sided Pyramid

Just as Imhotep, Pharaoh Djoser's vizier, was the creator of the stepped pyramid, Pharaoh Sneferu made the advances in construction that made it possible to obtain a smooth-sided pyramid at the beginning of the Fourth Dynasty. The pyramids were symbols of sun worship, identified with the perfection of the god Ra and with the sacred Benben stone. According to Edwards, both the pyramids and the Benben stone represent the sun's rays, thus symbolizing the immaterial becoming material (Edwards 1993: 281).

Without prejudice to religious interpretations, I believe that this transformation was the result of the search for a practical solution to prevent the build-up of sand on the steps. This flaw was an unpleasant sight that detracted from the perfection they sought to achieve in their works. The seemingly simple idea of laying an outer layer to smooth down the sides was not without its difficulties when it came to its practical implementation.

Unlike the stepped pyramids, in which measurement errors could be absorbed in each step, errors in the smooth-sided pyramid could not be hidden; they accumulated and were there in plain view. Sunlight and the shadows cast on the sides of the pyramid magnified these flaws. One of Pharaoh Sneferu's works at the beginning of the Fourth Dynasty was precisely to turn the stepped pyramid at Meidum built by Pharaoh Huni in the Third Dynasty into a smooth-sided pyramid. The plotting involved demarcating the entire pyramidal shape on the stepped structure.

The demarcation was undertaken by using strings fastened to props, which is the procedure usually used in setting out buildings on site. The blocks for the outer layer were then placed by following the shape made by the strings.

The existing stepped structure provided access to the apex, from which the plotting starts by locating the summit point. The summit is determined in such a way that the plotting contains the stepped structure and leaves room for the outer layer to be laid. It is from the summit point that the pyramidal shape begins to be plotted, the edges being projected downwards as far as the base.

Figure 6: The Pyramid at Meidum

It is often proposed that the pyramidal shape was drawn from the bottom up, i.e., from the ground to the top. The first difficulty lies in the impossibility of accurately plotting the square base around the stepped pyramid. Since it is not possible to see the opposite corners of the base or draw the diagonals, the plotting would be extremely inaccurate.

The outer layer was made up of casing stones carved in fine limestone in a trapezoidal shape, supported by the so-called backing stones (see Fig. 7). The space between the backing stones and the core structure was filled with the so-called packing stones. These blocks, like the core, were carved in poor limestone sourced from local quarries (Arnold 1991: 168).

"This separation of casing and core was crucial in pyramid building and determined, together with the step like shape of the casing construction, the structure of these buildings."

The collapse of the casing and part of the core of the pyramid at Meidum exposed the stepped core in the shape of a tower. The prevailing view among Egyptologists is that the pyramid was

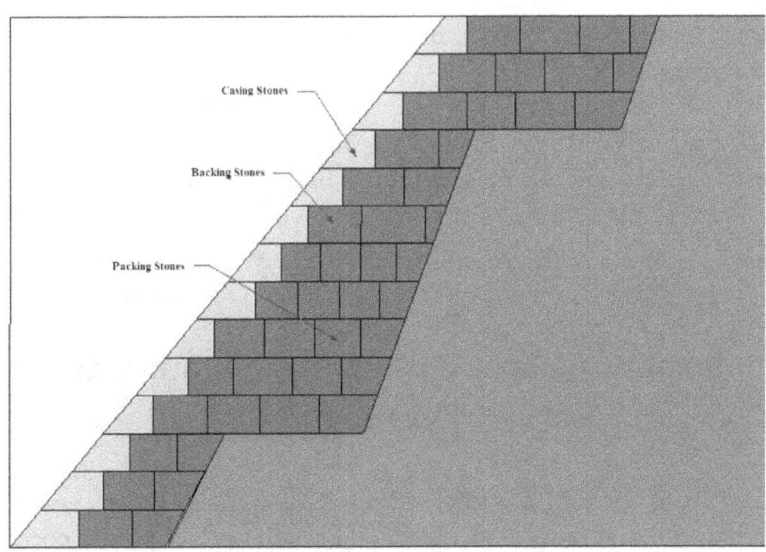

Figure 7: Structure of the casing

Figure 8: The Red Pyramid

dismantled and used as a stone quarry after the Old Kingdom. The purpose of this activity was to extract the fine white limestone, which is scarce in the area and was found in the successive outer layers of the pyramid. At present, the pyramid looks like a large tower with three steps in the middle of a mound of rubble and sand. Pharaoh Sneferu also built two pyramids at Dashur: the Bent Pyramid and the Red Pyramid.

The progress made by Sneferu at Dashur resulted in remarkable advances in the shape of the royal tombs as well as in the procedures to obtain stable structures which made it possible to build smooth-sided pyramids, the peak being reached with the Great Pyramid.

Structure of the Core in the Smooth-Sided Pyramids

The core of the smooth-sided pyramids is hidden under the casing that gave it its pyramidal shape. The difficulty of seeing the core has given rise to different interpretations regarding what the structure is like (Arnold 1991: 159). Initially, the opinion prevailed that the smooth-sided pyramids that began to be built in the Fourth Dynasty had a core made up of accumulated layers, like the Third Dynasty pyramids.

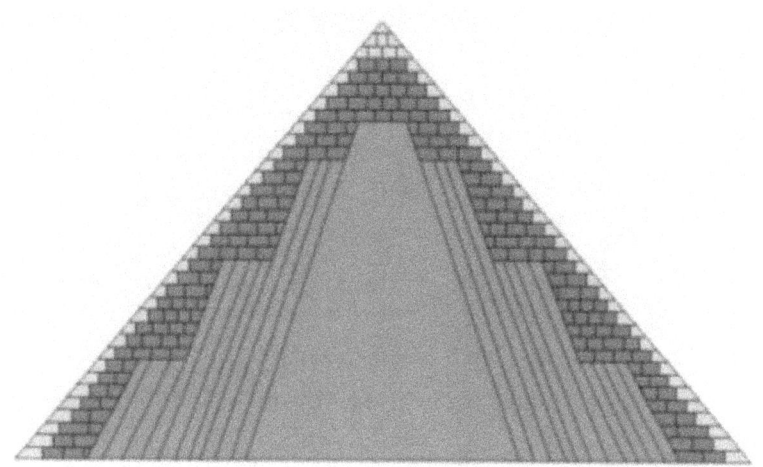

Figure 9: Diagram of the Pyramid of Meidum

Explorations by Maragioglio and Rinaldi in Giza found that the core of the Fourth Dynasty pyramids is made up of rows of horizontal blocks. They came to this conclusion after examining the tunnels dug by looters, in which the horizontal arrangement of the blocks can be seen (Maragioglio and Rinaldi 1965: 16) (Sampsell 2000).

The stepped pyramids of the Third Dynasty were built with leaning layers and small blocks, while the later ones, during the Fourth and Fifth Dynasty, were built with courses of larger horizontal blocks (Isler 1926: 121). The available archaeological evidence indicates that the shape of the core of the smooth-sided pyramids is stepped. The gash made in the pyramid of Menkaure (IV dynasty) in the year 1215 by Caliph Malek exposes a stepped core on which was laid the casing that gives it its pyramidal shape (Mendelssohn 1974: 115).

Martin Isler too refers to this evidence, "the gash left by the Mamluks on the north face of the pyramid of Menkaure has allowed researchers to identify at least three large steps inside the structure.

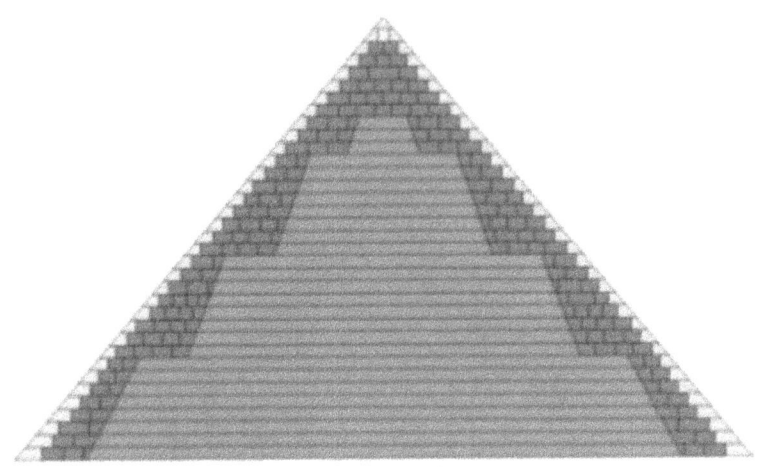

Figure 10: Diagram of the Smooth-Sided Pyramid

There is also a hint of four levels of a central core in the casing stone when observing the northeast corner of the pyramid of Khafre (Isler 1926: 192). In the satellite pyramids existing in Giza, the stepped core with which they were built is visible.

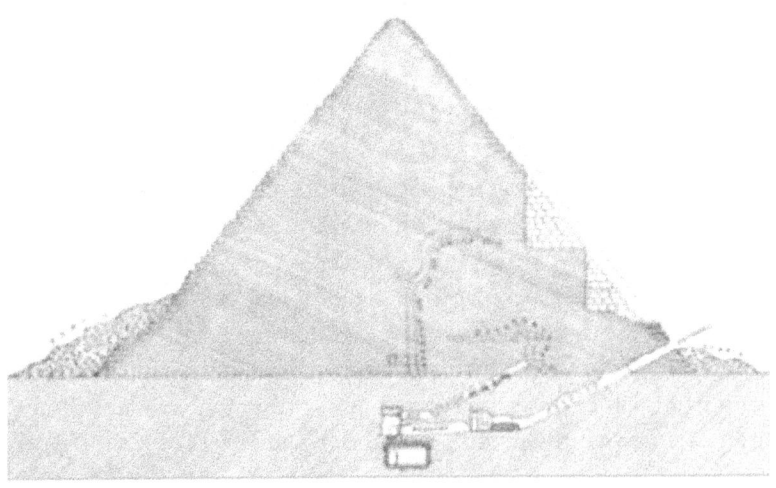

Figure 11: Sectional view of the Pyramid of Menkaure

Even in these smaller pyramids, in which the shape could be obtained with greater ease, a stepped core was built on which the casing was laid (see Fig. 14).

Figure 12: Satellite Pyramids in Giza

According to Dieter Arnold, since all the pyramids before and after the Fourth Dynasty were built with a stepped core, it is to be assumed that those of the Fourth Dynasty also have it, although it has not been sufficiently demonstrated in all of them (Arnold 1991: 168).

According to Kurt Mendelssohn: The good condition of the pyramids at Dashur (Fourth Dynasty) makes it impossible to see their core, as is also the case with the pyramids of Cheops and Khafre in Giza, "However, you can be sure that they were also designed in the same way" ... .., according to the archaeological evidence found in the existing gash in the pyramid of Menkaure (Mendelssohn 1974: 115).

The structure of the smooth-faced pyramid is a stepped core made up of horizontal courses, with large poor limestone blocks in the bottom sector which become smaller towards the top. The casing is composed of packing and backing stones that give it a pyramidal shape on which the casing blocks, made of fine white limestone, rest. According to Dieter Arnold, "There is no doubt that the casing, backing and packing stones were treated as a structural unit to be built simultaneously." (Arnold 1991: 82)

The Great Pyramid

Following Sneferu's construction breakthrough of building smooth-sided pyramids, Pharaoh Khufu had the Great Pyramid built. Building this masterpiece required strict compliance with the construction requirements, in the most remarkable landmark in the evolution of the pyramids. The erection of the Great Pyramid met two basic construction requirements.

Figure 13: The Great Pyramid

Construction Requirements

1. Building the tallest Pyramid. The builders' goal of erecting pyramids as tall as possible is clearly documented in the evolution of the pyramids.

The tallest pyramid in this evolution of construction is that of Pharaoh Cheops, with its 146 metres in height. This pyramid was stripped of its casing and its last 10 courses of blocks, so the original height is not known precisely.

2. Achieving perfection in shape:

The requirement of achieving perfection in the pyramidal shape was as significant as that of height, judging by the strict compliance with them.

In 1883, Flinders Petrie (the Father of Egyptian Archaeology) revealed the results of the measurement of the Giza pyramids in his book "The Pyramids and Temples of Gizeh." These results were confirmed with some minor differences by subsequent measurements performed by J. H. Cole (a surveyor with the Egyptian Ministry of Finance) in 1925. These differences are basically due to the fact that while Petrie used one point as a reference on each side of the pyramid, Cole used two. The book discusses the amazing accuracy achieved in plotting the Great Pyramid. Petrie's measurements give us the dimensions of the base as well as its orientation. It also leaves us with some unanswered questions:

Why did they plot the Great Pyramid so accurately and how did they do it? What was the original height of the Great Pyramid? Is there any relationship between the height of the pyramid and its base?

The accuracy achieved in plotting the base of the Great Pyramid can be summarized in this Petrie quotation: "So we must say that the mean errors of the base of the Great Pyramid were somewhat less than .6 inches, and 12" of angle."

The average length of each side of the square of the base is 230.347 metres. The base is rotated counter-clockwise by an estimated 3.5 minutes in relation to the cardinal points, or as much as 5', according to some measurements. This deviation is somewhat less than .1 degrees. We know that the human eye can detect a deviation of 1 degree. The rotation of the base has traditionally been regarded as the main error of the Egyptian surveyors. It is assumed that their intention was to orientate the pyramid according to the cardinal points and they managed to do it with that small error. One must consider that the human eye can detect up to 1 degree of deviation, and the deviation of the pyramid is 0,1 degrees.

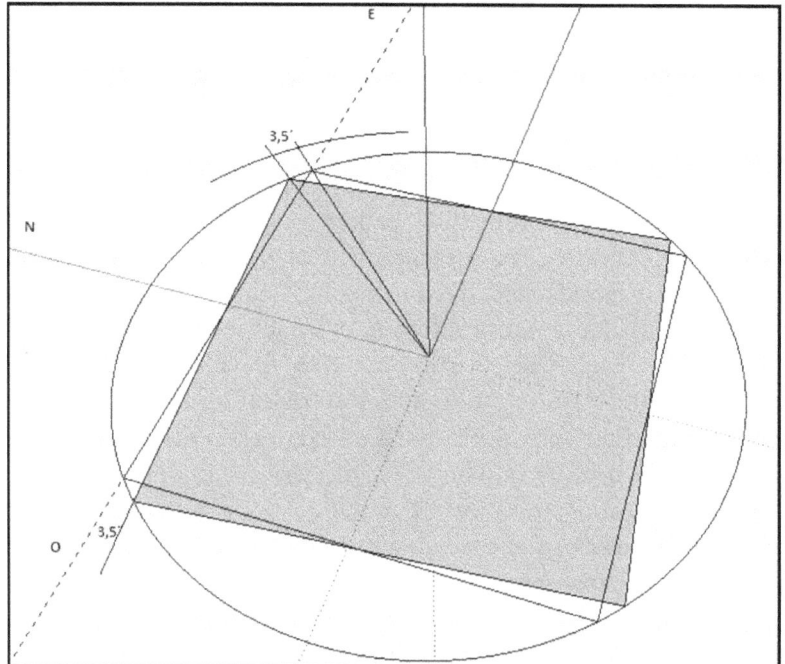

Figure 14: Rotated base of the Great Pyramid

The pyramid of Khafre exhibits amazing but also somewhat less accurate plotting, which Petrie summarized as: "the mean errors of the base of the Great Pyramid were somewhat less than .5 inches and 33" of angle." The average length of each side of the square of the base is 215.262 metres.

Interestingly enough, the square of the base in this pyramid is also rotated counter-clockwise by an angle of 5 to 6' of a degree. The so-called "error" in orientation that the ancient surveyors committed in the pyramid of Khafre is practically the same as that observed in the pyramid of Khufu.

Is this an error or was there some reason to orientate these large pyramids that way?

Finally, the measurement of the Pyramid of Menkaure yielded the following results: the mean errors of the base of the pyramid of Menkaure were somewhat less than 3 inches and 14' 3 "of angle.

As observed in these results, the accuracy of the smooth-sided pyramids increases with size.

Khufu side = 230.347 metres mean errors =0.6 inches

Khafre side = 215.262 metres mean errors = 1.5 inches

Menkaure side = 105.5 metres mean errors = 3 inches

Khafre's side length is twice that of Menkaure, but the mean error is half as much. In Khufu the mean error is also half that in Khafre, while the side length is slightly greater. We shall analyze this point in more detail when dealing with the plotting of the pyramid of Khufu.

The results obtained fill present-day surveyors with amazement. The ancient surveyors were aware of the perfection they had achieved, but they did not have the instruments to perform such measurements.

Accuracy or Perfection

Before dealing with the plotting of the Great Pyramid, let us analyze why the plotting should be so perfect.

From the pyramid of Meidum to the Great Pyramid, the Egyptian surveyors gained experience and acquired the skills that guaranteed the completion of plotting with perfection.

Figure 15: Casing Stone

First of all, we have to place ourselves in the context of the times. The ancient Egyptians used inaccurate measuring tools, so they never knew the average difference between the sides of the base of the Great Pyramid. This error is very small and was measured by Flinders Petrie early in the nineteenth century using instruments with optical least count.

The Pharaoh demanded perfection, rather than precision that he could not measure. He wanted a smooth-sided pyramid with no noticeable flaws. That was what he, like any other observer, could appreciate and judge. Unlike the stepped pyramids, in which the errors are absorbed by each step, the errors in the smooth-sided pyramids accumulate and can be seen clearly.

An example of that perfection demanded by the Pharaoh is the casing laid on the Great Pyramid, of which some blocks still survive.

Measurements made by Petrie reveal great precision in the surviving stones on the north face of the pyramid (1). The error between horizontal faces does not exceed 0,2 mm in each metre of length. The goal here is to seek perfection in the joints of the blocks.

Measurement between the faces of the blocks would have called for the use of a precision instrument .... that they just did not have. In fact, they never measured it accurately. They simply determined a height with the precision allowed by their measuring rod and then transferred it.

The instrument traditionally used to compare or transfer measurements is the stonemason's callipers. They consist of two curved metal legs joined by a hinge. These callipers are a very simple and ancient instrument which is used to transfer measurements with a precision of tenths of a millimetre. They could therefore transfer the measurement with a precision of tenths of a millimetre as many times as necessary, but they did not know how much it measured in tenths of a millimeter. The same thing happened with the plotting of the smooth-sided pyramids; they knew that the plotting was near perfection but they could not measure it.

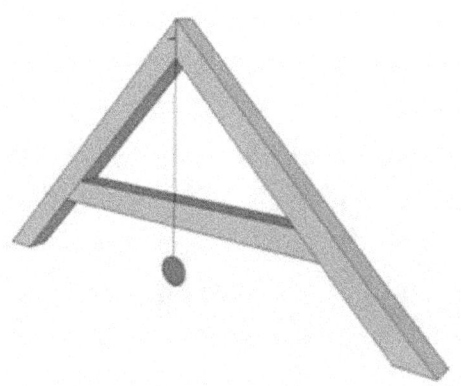

Figure 16: The square level

Levelling Work

The execution of a civil work begins with the surveying work. The survey includes setting out the work on site, which involves plotting the ground plan of the building to be erected. The setting out of the work requires leveling the ground on which the plotting will be undertaken. The ancient Egyptians had an instrument to control levels called the square level, and they could also resort to channeling water so as to get an accurate horizontal level. Surface to be leveled was 5.23 hectares, which is the surface occupied by the base of the Great Pyramid. However, of this surface, only the perimeter on which the square of the base of the pyramid was plotted, the pavement and the surrounding area were leveled. There is a mound at the base of the pyramid about 15 meters high on which the pyramid was built. This mound is part of the plateau itself and is visible from the shaft that was dug inside the pyramid connecting the entrance to the Grand Gallery with the descending passage (see arrow Fig: 17).

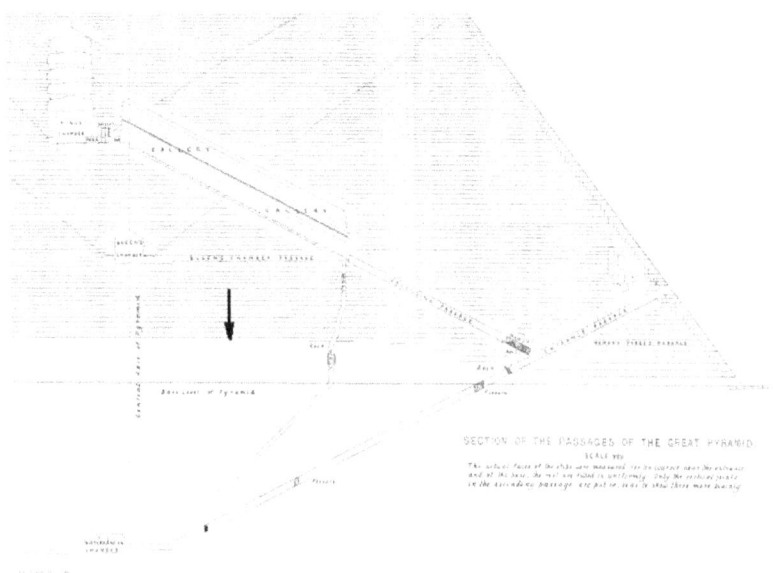

Figure 17: Sectional view of the Great Pyramid

Figure 18: Base of the Pyramid of Khafre.

As noted above, the plotting of the square for the base of the pyramid was done with amazing accuracy and precision. If this plotting had been performed by measuring on the ground, then the ground would have been leveled to draw the diagonals of the square. The diagonals were indispensable both for drawing the square and ascertaining its accuracy. The existence of this mound indicates that the base of the pyramid was plotted around it and without being able to draw the diagonals. Another difficulty is the inability to plot this large base with the accuracy with which they did without the availability of precision instruments.

This is one more reason to think that they did not measure directly on the ground but used a different technique that allowed them to plot the base accurately. The same thing happens with the base of the second great pyramid, the pyramid of Khafre. The land was not levelled, and a sector of the plateau itself is part of the pyramid. This mound is clearly visible from outside as a result of the casing that hid it having been removed (see arrows Fig. 18 and 19). The original level of the plateau can be seen in the left sector. The space between this sector and the pyramid was excavated, and there are traces of the quarry work done in the place where stone blocks were extracted. On the right in the lower sector of the pyramid, you can see the continuation of the plateau which was excavated into steps to support the casing. On the other hand, even though the ancient Egyptians had the labor force of an entire kingdom at their disposal, they did not work unnecessarily. They spared themselves the trouble of leveling the ground and cutting, transporting and placing the blocks that would have taken the place of that mound. They also benefited from the existence of a central sector with great structural stability.

Figure 19: Mound at the base

The largest sundials are the most accurate, just as the largest pyramids are the most accurate in their plotting.

Daniel Gerardo

DANIEL GERARDO

CHAPTER II

Construction Stages

Like all civil works, building the great pyramids of Giza required dealing with their survey before proceeding with the construction work. Erecting any building without surveying it first is a hopeless task, and all the more so in the case of a smooth-sided pyramid, because the plotting of a smooth-sided pyramid determines the stages of its construction.

The plotting of a smooth-sided pyramid is different from that of any other building, precisely because of its pyramidal shape. In the case of a building, for example, the ground plan is plotted on the levelled ground; this will be the base of the building. Then, the edges of the building are vertical and plotted as construction work progresses by using, for instance, a simple plumb line.

In the case of a smooth-sided pyramid, both plotting and construction were approached in a different way. For example, the surface of the ground under the pyramids of Cheops and Khafre were not levelled, which indicates that they did not survey the square for the base either, as it was impossible to draw the diagonals or see the opposite corner from any of the corners, as explained above.

When plotting a smooth-sided pyramid, the edges had to be straight and converge on the summit point 146 metres off the ground (the height of a 48-story building). The edges cannot have been plotted as construction work progressed because, again, it would have required the use of precision instruments and the necessary surveying know-how.

"Careful surveying during construction was essential. Otherwise a twist might occur and the edges would not meet at a point at the top." (Hawass, 1)

Surveying the pyramidal shape as the pyramid was built would have resulted in snowballing errors which would have prevented the four triangular sides from meeting at the summit point. Let us add that once these inevitable errors had been committed, it would have been impossible to correct them, since they would have become noticeable with construction at an advanced stage, close to the summit.

In the case of Meidum, the casing was laid over a very inaccurate core. Everything points to remeasuring after construction of the core, before laying the casing. Something similar can be observed in the Pyramid of Cheops, in which the core looks somewhat misaligned with the final shape of the casing. The courses are not perfectly horizontal, as they should be if the pyramidal shape had been plotted from them. (Isler 1926: 210).

Unlike the stepped pyramids, in which errors can be absorbed in each step, the edges of the smooth-sided pyramids are supposed to be straight and meet at the top. Any deviation or warp in an edge would be clearly visible. For a pyramid the size of the Pyramid of Cheops, a deviation of 2 degrees at the base grows to 15 metres at the top, so the edges would not meet at the summit point. Plotting the pyramidal shape without determining the summit point would require the use of precision instruments that the ancient Egyptians were certainly not familiar with. First, it is necessary to build the stepped core to establish the location of the summit point, which is essential for plotting the pyramidal shape from it (Mendelssohn 1974: 116).

Accuracy of the smooth-sided pyramids lies in their casing, which was laid over an inaccurate stepped core that was constructed first. This is logical and usual in all constructions; completion is achieved

Figure 20: "One-Stage" Construction

by laying the casing over a structure that has been built first. The traditionally accepted way of plotting the pyramid – from the bottom up, or from the ground – is a misconception that we have incorporated and is the source of much of the confusion. This leads to a false scenario in which it is possible to build the pyramid in one single stage, all at once.

In their quest to solve the lifting of blocks and making a more efficient ramp than that first proposed by Ludwig Borchardt, the researchers are on a wrong track that the ancient Egyptians were never on and does not lead to the Great Pyramid. This creates a maze of theories that seek to solve unheard-of problems that the ancient Egyptians would not have understood and from which it is only possible to escape by plotting the pyramid the way the ancient Egyptians did.

The scenario of the Egyptian surveyors, and Pharaoh Sneferu in particular, was a very different one. He simply considered how to turn the stepped pyramid of Meidum into a smooth-sided pyramid. It was then a matter of plotting the pyramidal shape over an existing stepped pyramid. The pyramidal shape was plotted from the summit point down. This problem is altogether different from plotting the pyramid from the ground up, and it is the one that the ancient Egyptians solved.

Figure 21: Construction Stages

The existence of the stepped core inside the smooth-sided pyramids, as we saw in the previous chapter, confirms that there were no changes in the way pyramids were built. They would begin by constructing the stepped core, over which they would then plot the pyramidal shape. Plotting the Great Pyramid from the bottom up as the pyramid was built, as well as building the pyramid in a single stage, was a problem with no solution at the time. What was within their power – and what they did – was to plot the pyramidal shape from the summit point and over the stepped core they had already built.

Building the Stepped Core

The erection of every smooth-sided pyramid began with the plotting and construction of an inaccurate stepped core. The core accounted for the bulk of the volume of the pyramid, and its construction was undertaken with straight ramps with a wide surface in the lower and middle sectors, as well as ramps resting on the steps of the core, in the upper sector. Multiple straight ramps were used in the lower sector, since in the case of the first courses, the ramps were simple embankments with a wide surface. The lower sector is where the largest volume of the pyramid is and where the largest numbers of workers were used.

The ramp from the quarry reached between medium height and as high as 70 metres. At this height, large granite blocks were placed such as those in the roof of the Upper Chamber (60 tonnes). This ramp went from the quarry to the W-S corner of the stepped core. Beyond that point, the ramp rested on the ground and the stepped core on the west, north and east faces, reaching a height of 70 metres.

Vestiges of ramps this high (70 m) were reported by Borchardt in 1920 at the pyramid of Meidum. On the basis of these discoveries, he proposed the use of large straight ramps to build smooth-sided pyramids. This was the first proposal regarding the construction of the Great Pyramid, and one that led to many objections and a large number of theories proposing different kinds of ramps. This is due to the fact that the Great Pyramid is twice as tall as the pyramid at Meidum. The volume of material to be accumulated on the ramp increases exponentially with height, and in the case of the Great Pyramid, it would exceed that of the pyramid itself.

Mark Lehner argues that the straight ramp cannot have gone from the quarry to the top of the pyramid because it would have had too steep a slope. According to Hawass, Lehner located the quarry on the south side of the Pyramid of Cheops. That is the only direction that could have accommodated the ramp. Towards the east and west, there are tombs from Cheops' reign, while towards the north there are no traces of quarries, and the pyramid is near the edge of the plateau (Hawass). (3)

The ramp started at the mouth of the quarry and extended for about 32 m as far as the south-west corner of the pyramid with a total height increase of 37 m and a slope of about 6 degrees 36 minutes. A ramp of similar dimensions is described in Papyrus Anastasi I (late New Kingdom). Recent discoveries reported by Zahi Hawass confirm the existence of traces of this straight ramp that went from the quarry located in the south of the Giza plateau to the South-West corner of the pyramid.

During construction of the lower and middle sectors of the core, it was necessary to move a significant number of large blocks. There was ample room for maneuvering these blocks into place using vast numbers of workers.

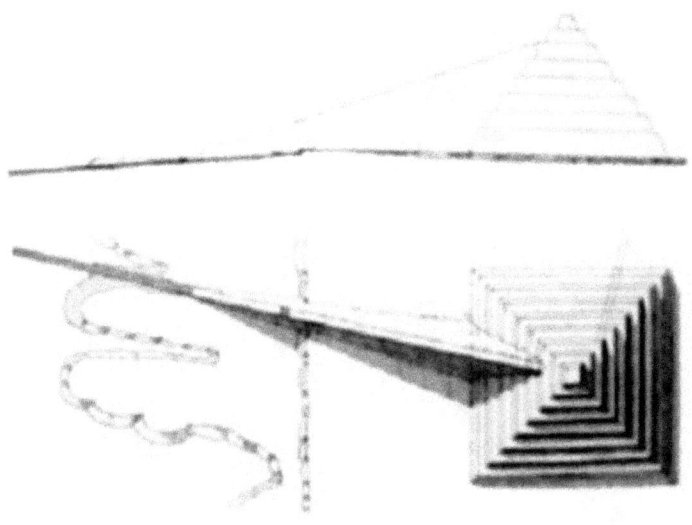

Figure 22: Ramp used for the Pyramid of Meidum

Considering the plotting of the ramp and the presence of large blocks (60 tonnes) in the roofing of the King's Chamber at 68 metres, the rest of the ramp must have rested on the steps of the core on the west and north faces, reaching that height.

The granite slabs in the King's Chamber came from the distant quarry at Aswan and were brought by barges to the harbour in Giza. An accessory ramp connected the harbour with the main ramp coming from the quarry, on which these slabs were raised to their final position. Beyond the height of the large blocks in the roof of the King's Chamber (68 m), construction of the structure carried on with the traditional method, that is, by using ramps resting on the steps of the core.

Figure 23: Mark left by the ramp on the pyramid at Meidum

The lower and middle sectors are where the bulk of the volume of the core and the larger blocks that required large crews are concentrated. The size of the blocks in the core decreases with height, which is evidence that the difficulty of moving them up increased with height as room on the ramps and on the structure decreased. Largest number of workers would come during the flood months, and the ramps had to be broad enough to accommodate a large number of crews at work.

A common misconception is that if there are 2,300,000 blocks in the Great Pyramid and it was built in 20 years by working an average of 10 hours a day, then it was necessary to fit one block every 2 minutes, which is considered impossible. It is impossible if one crew worked alone ... but the work involved hundreds of crews, and even more of them during the floods. If we consider 300 crews working on raising the blocks to the course under construction, each crew had an estimated 10 hours (2 minutes x 300 crews) to put each block into place.

Another misconception is the assumption that the Great Pyramid was built differently from the other smooth-sided pyramids. The Great Pyramid's height increase did not imply any changes to the construction techniques implemented up until then. The spiral ramp resting on the steps made it possible to build that stepped core without any trouble. The only new difficulty posed by the great height is laying the casing on the upper sector, which we shall analyze later. Once the core was completed, any ramps hampering the plotting were removed, and the pyramidal shape was then plotted using strings attached to pegs.

Plotting and Orientating the Casing
Thales's Theorem:

We will now begin to analyze the plotting of the Giza pyramids. We will resort to the reasoning of an ancient thinker who determined the height of the Great Pyramid (Thales of Miletus). We will also cite the reasoning of a modern surveyor and archaeologist who worked out the height of the Great Pyramid (Flinders Petrie). It is interesting to compare and see how each of them reasons depending on the knowledge and possibilities at the time.

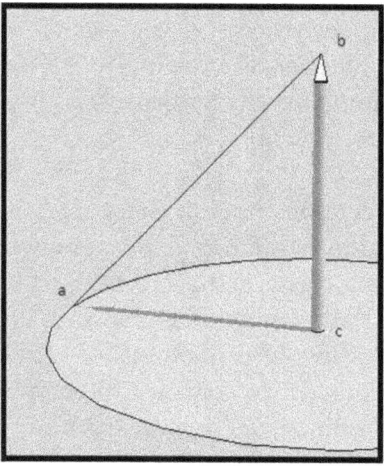

Figure 24: Model

The Greek mathematician Thales of Miletus (620 to 546 BC) determined the height of the Great Pyramid without measuring it directly, according to the account provided by Plutarch.

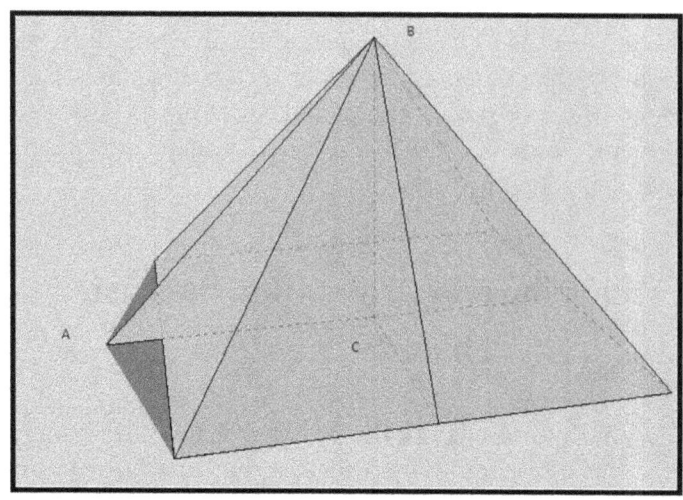

Figure 25: Shadow Cast by the Pyramid

He used one of the theorems named after him. His reasoning was, "My shadow relates to me exactly as the shadow of the pyramid relates to the pyramid itself." From this, he deduced that when his shadow was equal to his height, the shadow of the pyramid would be equal to its height. To apply this theorem, Thales drew a circle with its radius equal to his own height.

Then he placed in the centre of the circle a pole (gnomon) whose height was equal to his own height, and waited for its shadow to reach the circle. The moment the shadow touched the circle, he signaled to his assistant, who marked the point of the shadow of the pyramid with a stake. He then measured the length of the pyramid's shadow as marked, which had to be equal to the height of the pyramid. Triangle ABC, formed by the shadow of the pyramid, is equivalent to triangle abc, formed by the shadow of the model, because they have the same angles. This is so because both triangles are equivalent in that they have the same angles and their sides are proportional. In this case the rays are incident at a 45° angle, so that both sides of each triangle are equal. The shadow cast by model "ab" is equal to the height of pole or gnomon "cb," while the shadow cast by pyramid "AC" is equal to its "CB" height.

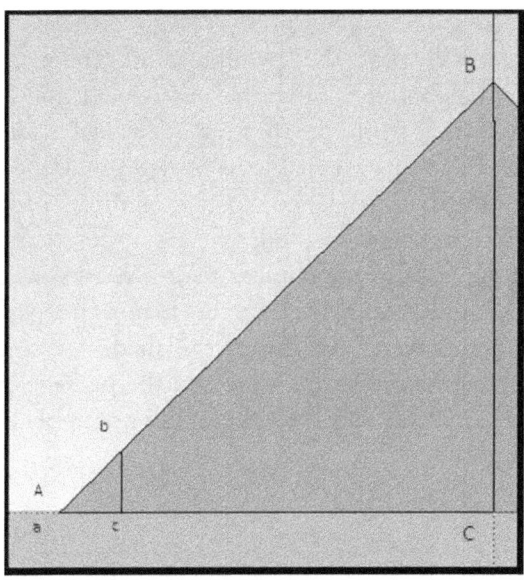

Figure 26: Thales's theorem

This determination of the height of the pyramid is an estimate. The shadow cast by the top of the pyramid on the ground is not very sharp, and the point where the shadow ends cannot be determined accurately. The sun is not a point source of light; as seen from Earth, it has an angular size of an estimated .5 degrees, which results in an area of penumbra between sunlight and shade. For the height of the pyramid, the width of this penumbra is approximately 1.10 metres.

We can then place triangle abc inside triangle ABC, whereby the equivalence and proportionality become more noticeable. Flinders Petrie's determination of the height of the Great Pyramid was an approximation too. Because the top of the pyramid had been removed, as had been the last 10 metres of the course of blocks, the figure that Petrie came up with is also an estimated height. He calculated this height using the size of the base and the slope of some casing stones that were still in place.

"In general, we probably cannot do better than take 51° 52 '± 2' as the closest approximation to the mean angle of the pyramid. The mean base being 9,068.8 ± 0.5 inches (230.347 metres), this yields a height of 5,776.0 ± 7.0 inches (146.71 mt)." Petrie.

In chapter we will plot the pyramids of Cheops, Khafre and Menkaure. The heights of the pyramids of Khafre and Menkaure will match those we get from the plotting. This is evidence that the plotting is correct, and it will enable us to determine the exact original height of the Pyramid of Cheops. The plotting of the casing is relatively simple and intuitive, but we are used to measuring, and these old techniques are not familiar to us. Works in ancient Egypt were relatively simple and there were no major changes over time to the techniques employed. The use of the shadow cast by a gnomon was the basic tool for measuring time and the passage of the seasons and to plot the pyramids. This technique did not apply to the plotting of other buildings.

The gnomon's curved shadow paths

In the course of each day a gnomon will produce a curved shadow path. In the course of a year, these curved paths will be found between the curves of the winter and summer solstices. The summer solstice marks the highest position of the sun in the sky, while the winter solstice indicates the lowest position of the sun in the course of the year. In addition to the Earth's rotation and translational motion, there is also the sun's declination, caused by the shifting position of the Earth's axis. The sun's declination movement is continuous throughout the year, except on the solstice days, when there is no declination. The sun's declination is noticeable in that the sun rises at a different point in the east and sets at a different point in the west every day. The sun's path on the solstices is east-west, and there is no declination. For this reason it is always suggested that the pyramids were orientated during the summer solstice. On the equinoxes, the sun rises at the point closest to the cardinal point east, and it sets at the point closest to the cardinal point west. On the equinox, the movement of the sun will have a declination, just as it will have any other day for the rest of the year.

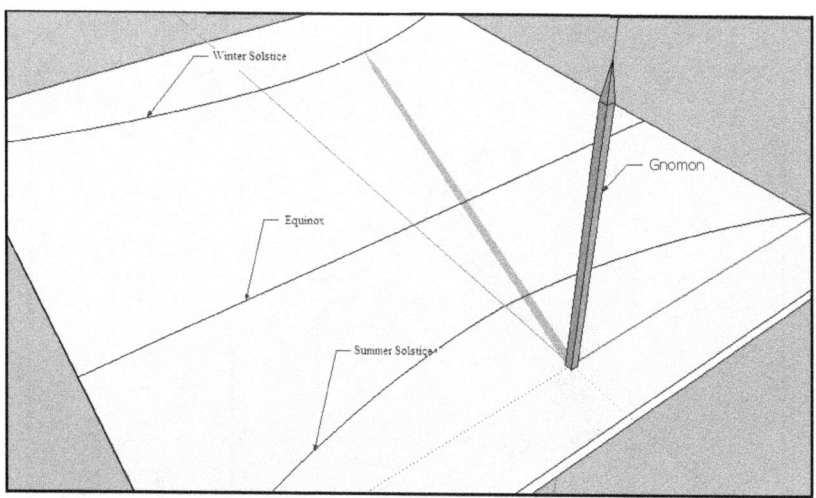

Figure 27: Shadows cast by the gnomon

Characteristics of the plotting of the pyramids of Giza:

The Great Pyramid acts like a gnomon, with its summit casting its curved shadow path on a daily basis. The size and plotting of the square base of the pyramid were calculated in such a way that the corners on the north side, points 1 and 2, touch the curved shadow path of October 11th. In the course of that day the shadow enters the pyramid by the N-W corner and exits it by the N-E corner. The same thing happens with the pyramid of Khafre, but on October 8th.On the morning of that day, when the sun reaches point 1 (see Figure), the sun's rays strike at the same angle of slope as the pyramid's edge, and the shadow of the pyramid's summit falls exactly on the N-W corner, point 1. The curved shadow path and the pyramid's N-W corner touch each other at that moment.

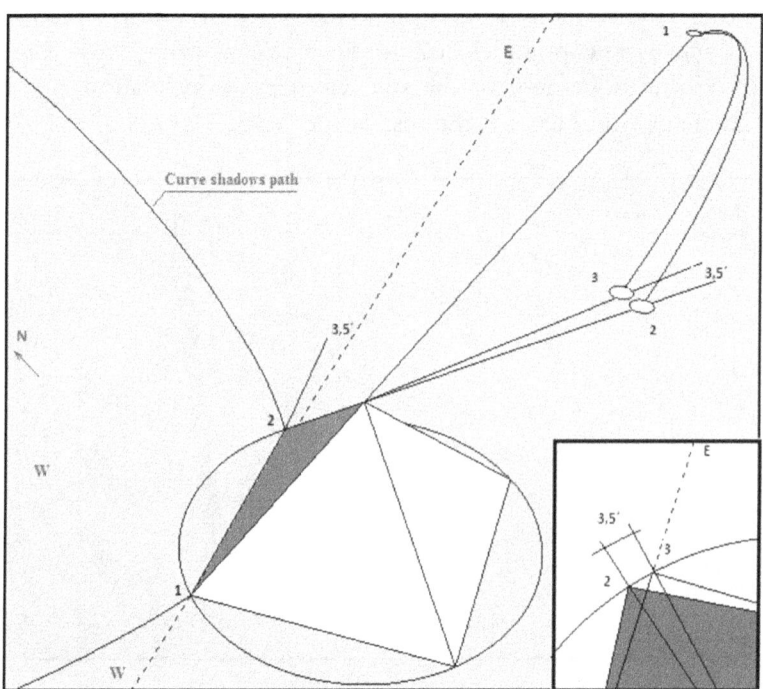

Figure 28: The Sun's Declination

The curved shadow path and the pyramid's N-E corner touch each other at that moment. As we explained above, the curved shadow path deviates somewhat towards the north due to the sun's declination, and the chord that joins two points on this curve will also be tilted towards the north.

By drawing a straight line joining both corners, points 1 and 2, we get the north side of the base of the pyramid, which is also the chord of the curved shadow path.

The side thus obtained will not have an E-W orientation, but will deviate slightly towards the north, as will the chord of the curved shadow path. This deviation is due to the sun's declination occurring between points 1 and 2, which can be seen in the angle between points 2 and 3. Point 3 in the sky is where the sun would be if there was no sun declination. Point 2 in the sky is the actual position of the sun with sun declination. The sun's average hourly declination during the month of October is an estimated 0.9 degrees. In the estimated 4 hours that the sun takes to go from position 1 to position 2, the sun's declination is 3.5 '.

The relation between the height of the pyramid and the side of the base is not accidental, but must clearly have been determined by plotting on a scale model. The curved shadow path is determined by the height of the pyramid and the day on which plotting is undertaken. The length of the side of the base must be such that it makes it possible to intersect the curved shadow path with the vertices of the north side of the square of the base. In addition, the length of the side of the base must be such that it allows the perpendicular to the midpoint of the north side to run through the centre of the base of the pyramid, and distance d must be half the side.

It must also be noted that since the curved shadow path is not symmetric, if the square for the base was plotted in an east-west direction, only one vertex could intersect it, the other vertex being some distance from the curved shadow path (25 cm). For both vertices to intersect the curved shadow path, not only is it necessary for the sides to be the right length but the square must also be rotated by an angle equal to the sun's declination with the axis of the pyramid as its centre. The rotation is counter-clockwise because the plotting was undertaken in October with the sun declining towards the winter solstice.

The shadow cast by the pyramid's summit allows us to plot the north side of the base of the pyramid as well as the edges of the face and its apothem. The rotation of the base of the pyramid is observed in the bottom right-hand corner of the figure.

In my view, this is the conceptual origin of the "Indian Circle" method proposed by Martin Isler regarding the orientation of the Great Pyramid. The pyramid itself is the gnomon, and its shadow – cast in accordance with Thales' theorem – was used to plot the pyramid.

When working on the model, they drew a scale circle on levelled ground with a diameter equal to the estimated diagonal of the base of the model. In the centre of the circle is the gnomon with the scale height of the pyramid. By joining with a straight line the points where the circle intersects the curved shadow path, they determined the chord, which is the side of the base. Next, they measured distance d on the perpendicular of the chord to the centre of the circle, which must be half the side obtained.

The relation between the height of the pyramid and the side of the base is unique and has to be determined by means of scale plotting. The casing of the pyramid is then plotted by casting the shadow of the model obtained.

In a 1/100 scale model the deviation is 2.5 mm, while at the base of the pyramid it is 25 cm.

Start of Plotting

The plotting of the pyramidal shape – such as in the pyramid at Meidum – began on the stepped core already built. The ground was then levelled within the perimeter around the core, which supports the casing. The pyramid's uppermost piece, called pyramidion, is small, and plotting it is like plotting a very small pyramid. The pyramidion is not plotted by measuring. Due to its small size, and for greater precision, the measurements are transferred. In addition, all its dimensions, the diagonals of the base, the edges, the height and – as we will see later on – the apothems which were also plotted are accessible. A rod is used for each of these dimensions. Once the pyramidion has been placed atop the stepped core, the summit point – from which the plotting will be undertaken by means of strings – is defined. The summit point is determined in such a way that if the pyramidion's edges are projected visually, there is enough room for laying the casing over the core.

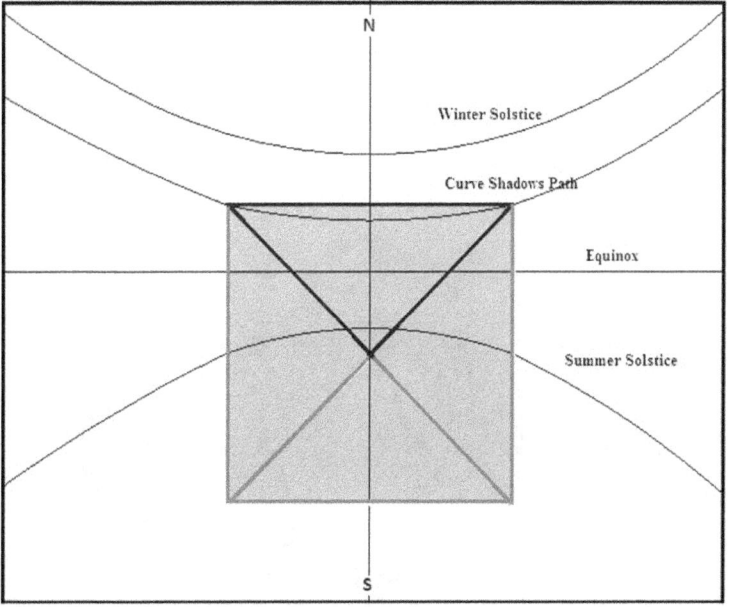

Figure 29: Plotting from the Summit

Once the pyramidion has been leveled, it is orientated in such a way that at some point on October 11th, the shadow will enter by the N-W corner and exit by the N-E corner.

The pyramidion is a small model, a small pyramid equivalent to the Great Pyramid because their angles are the same and their dimensions are proportional (see Thales' theorem).

The right scale for the model may have been 1/100, its dimensions being 100 times smaller than the Great Pyramid to be plotted. This model makes it possible to plot the casing of the Great Pyramid in accordance with Thales' theorem. The shadows cast by the pyramidion are equivalent and proportional to those cast by the pyramid, and are used to plot it. The first step of the procedure involves accurately plotting the pyramidion and projecting it towards the base of the pyramid.

The shadow model

The solar shadows cast by the pyramidion must meet the requirement of good sharpness and resolution. To improve the resolution of the shadows, we will use a shadow model rather than the pyramidion.

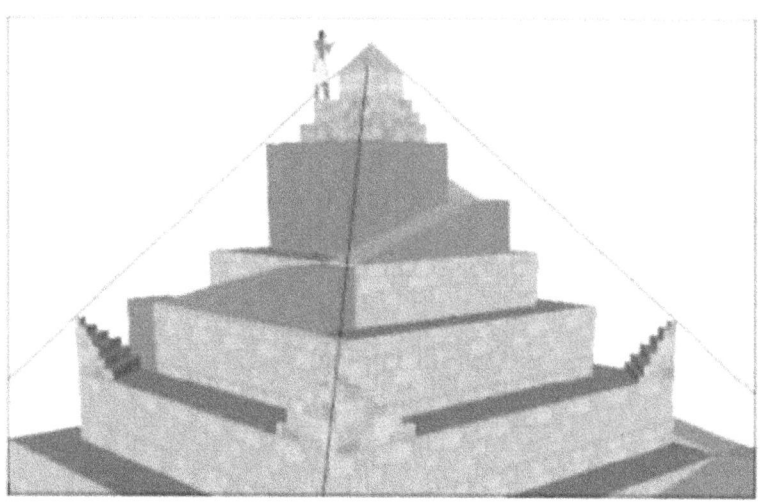

Figure 30: The Edges form the top.

The model will consist of a square base made of fine wood and a gnomon placed in its centre. On the shadow model, the shadow cast by the gnomon is seen along its daily path. The gnomon will have a sharper tip than the pyramidion to improve the resolution of the shadow cast. The resolution of this model is similar to that of a sundial. The resolution of a timepiece this size is 1 minute of time, which is equivalent to 145 mm. The largest known sundial has a resolution of 15 seconds. The larger the timepiece, the greater the distance between hours and the better its resolution. In the case of both sundials and the pyramids, the larger their size the greater their precision. Although it seemed rather illogical when thinking of plotting the pyramids by measuring, it is now clear that when you measure time with a gnomon, that is just what happens.

The largest sundials are the most accurate, just as the largest pyramids are the most accurate in their plotting.

Figure 31: Example of pyramidion.

Figure 32: The Model.

Sharpness of the shadow cast

The shadow cast by the gnomon on the basis of the model is sharp enough, as discussed above. However, the shadow cast by the gnomon towards the base of the pyramid, from a height of over 146 meters, will hit the ground without much sharpness. It is necessary for the shadow cast to be about as sharp as that cast on the model. This sharpness is obtained by transferring the shadow to the base of the pyramid by means of markers or gnomons. These gnomons are placed on the surface of the edge of the corner to be plotted on the north side, on the steps of the pyramid core. In the course of the morning the model will cast its shadow on the corner of the first step, thus determining the location of the first gnomon. This gnomon will cast a new shadow on the next step, thus determining the location of the next gnomon, and so forth all the way down to the base. A couple of weeks are available for this task, which involves determining on a daily basis the exact position of the plane containing the edge to be plotted. The location of the exact position of the corner of the base is determined the day the plotting was undertaken.

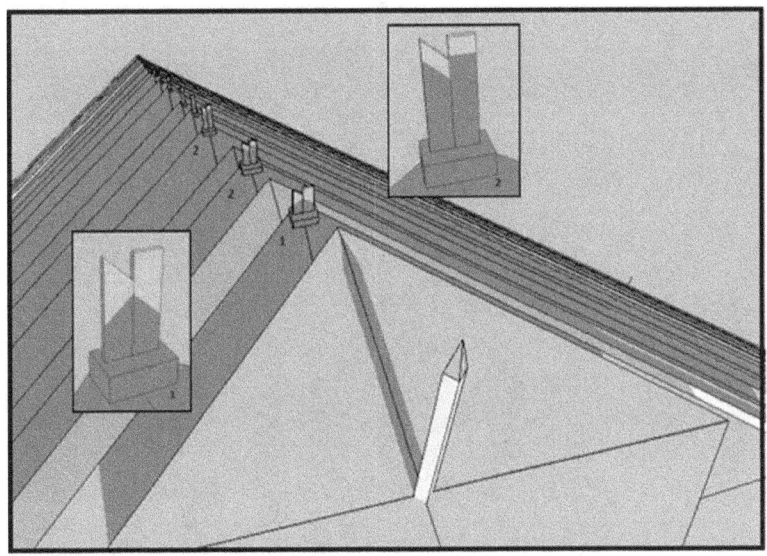

Figure 33: Transferring the shadow from the top.

That day, the sun's angular height coincides with the angle of the edge of the pyramid. The design of these markers or gnomons was probably just wooden strips with markings on one side and a notch in its upper end that only covered half the strip. *

Each strip, placed in a groove on a stone base that keeps them perfectly vertical, does a very good job of acting as a marker. For some days before the edge is marked out, the shadow cast by each wooden strip will be measured on the markings of the next wooden strip. This measurement is used for making sure that the wooden strips are placed equidistantly. All the strips will have their notches in the same position; in this case on the left, except for the last one, which is inverted. The day on which the sun's angular height coincides with the edge to be plotted, sunlight will shine through all the notches until it reaches the last inverted strip. A team of assistants – each of them next to one wooden strip – will turn each strip, starting with the one at the bottom, and check that it receives sunlight in the correct position. This manoeuvre has to be quick because the shadow moves quickly too. The positioning of the wooden strips on the previous days by using the shadow falling on

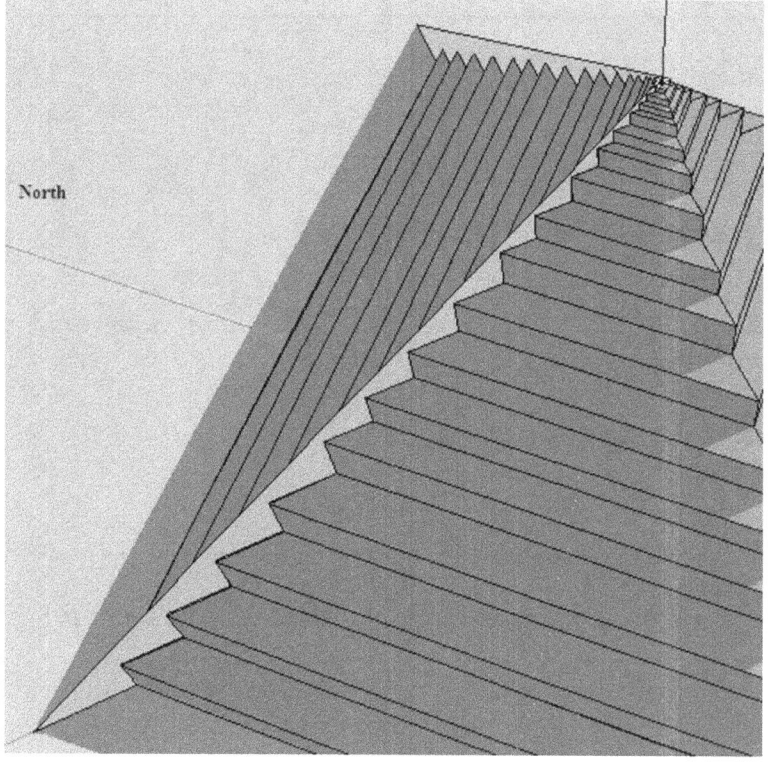

Figure 34: Shadows of the Edges.

e markings provides a preview of the final result. Once the locations of the corners of the north side of the base and those of the edges along their entire length have been determined, the wooden strips are replaced with pegs, and a string is placed that will mark the position of the edge. A piece of fabric that drops vertically and is fastened to the core of the pyramid is added to the string. When the edge is marked out this way, the string and fabric act as a screen that casts shadows and makes it possible to see shadows cast on it. One of the control checks to be made involves making sure that the edge does not cast any shadows when it should not, that is, when the sun is positioned in the direction of the edge. If any shadow is noticeable along the entire length of the edge, however small, it means there is an error to be rectified.

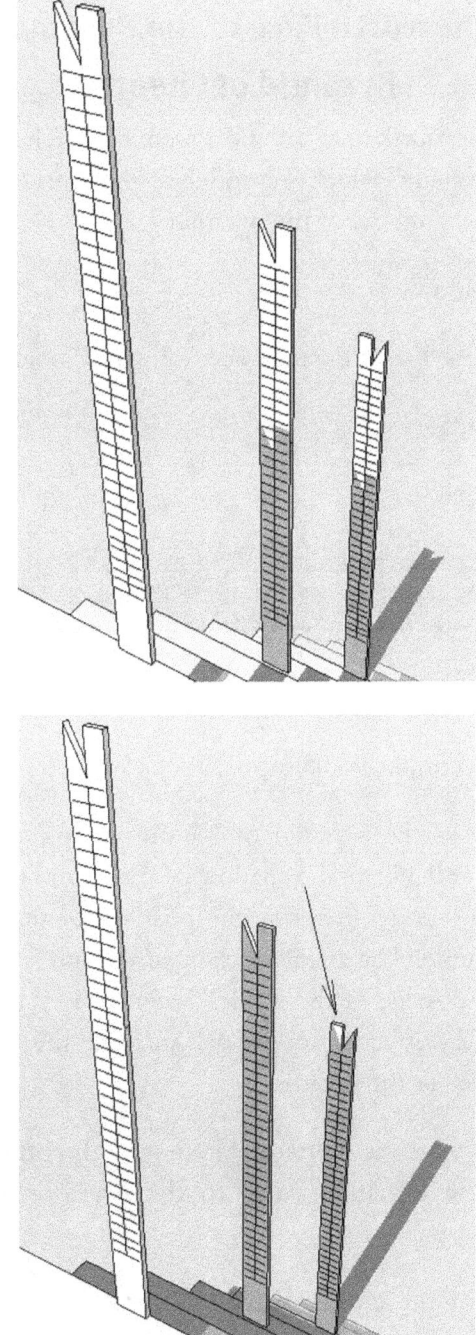

Figure 35: The Wooden Strips

Plotting the North Face of the Pyramids of Giza
Pyramid of Cheops

We will plot the north face of the Pyramid of Khufu by using an architecture program which reproduces the curved path of the shadow cast by the model of the pyramid.

Results of the plotting

Coordinates of the Great Pyramid according to Google Maps:

Latitude: 29.979254N

Longitude: 31.134201E

Side of the base of the pyramid: 230,34 m

Day: 11th October   9:36 am shadow on N-W corner

1:50 pm shadow on N-E corner

Height of the pyramid: 146,08 m

The north side of the pyramid of Khufu was plotted on October 11th. That day, at 9:36 am, the shadow of the gnomon falls on the N-W corner of the model, marking the position of that corner in the base of the pyramid. The angular height of the sun is the same as the slope of the N-W edge of the model. Later, at 1:50 pm, the shadow falls on the N-E corner of the model, marking the position of that corner in the base of the pyramid.

The north side, which is plotted by joining both corners, will have a counter-clockwise deviation equal to the deviation of the curved shadow path that day.

The deviation of the curved shadow path is a result of the sun´s declination, which is 0.95 minutes/hour on October 11th. The time the shadow cast took to go from the first to the second corner was 4 hours and 14 minutes = 4.23 hours.

The deviation of the curved shadow path, and therefore that of the north side plotted, will be (0.95 minutes/hour) x 4.23 hours = 4 minutes. It is the angle of rotation of the pyramid´s base.

The counter-clockwise rotation is due to the fact that the plotting occurred in October, with the sun moving away from the summer solstice. If the rotation were clockwise the plotting would have been undertaken in March.

As a result of this plotting, we conclude that the original height of the Great Pyramid was 146.08 metres +- 0.01, because it is the only height that makes it possible to meet the plotting conditions.

Pyramid of Khafre

Coordinates of the Khafre Pyramid according to Google Maps:

Latitude: 29.975643N

Longitude: 31.131455E

Side of the base of the model: 215,26 m

Day: 8th October   9:38 am shadow on N-W corner

1:47 pm shadow on N-E corner

Height of the pyramid: 143,62 metres

The north side of the pyramid of Khafre was plotted on October 8th. That day at 9:38 am the shadow of the gnomon falls on the N-W corner of the model, marking the position of that corner in the base of the pyramid. The angular height of the sun is the same as the slope of the N-W edge of the model. Later, at 1:47 pm, the shadow falls on the N-E corner of the model, marking the position of that corner in the base of the pyramid.

The north side, which is plotted by joining both corners, will have a counter-clockwise deviation equal to the deviation of the curved shadow path that day

The deviation of the curved shadow path is a result of the sun´s declination, which is 0.975 minutes/hour on October 8th. The time the shadow cast took to go from the first to the second corner was 4 hours and 9 minutes = 4.15 hours.

The deviation of the curved shadow path, and therefore that of the north side plotted, will be (0.97 minutes/hour) x 4.15 hours = 4 minutes.

Pyramid of Menkaure

The difficulty with the pyramid of Menkaure is that its casing was not completed, which makes accurate measurements impossible. Nor was it removed, as was the case with the Great Pyramid, in which Petrie was able to identify the sockets and use them to determine the precise measurements of the base.

This pyramid is smaller, so the error in the plotting will be greater and is present in the rotation of the base. The clockwise rotation is a sign that the plotting was undertaken in March.

Concavity of the Faces

The base of the Great Pyramid is not square, as it actually has eight sides forming eight faces. The apothems do not end on the sides but are recessed towards the centre of each face. Managing this concavity involved significant additional work. The concavity makes it possible to see the shadow cast by each edge on the apothem and the edge of the face containing them. When the shadow of one edge is cast on another edge, it marks out the side of the base containing them.

The shadows cast by these long screens on other screens of similar length have good resolution. However, the use of wooden strips with markings to improve resolution is not ruled out. The apothems have a steeper slope, which contributes to greater straightness of the string used to mark them out. The straightness of the apothems makes it possible to verify the straightness of the edges.

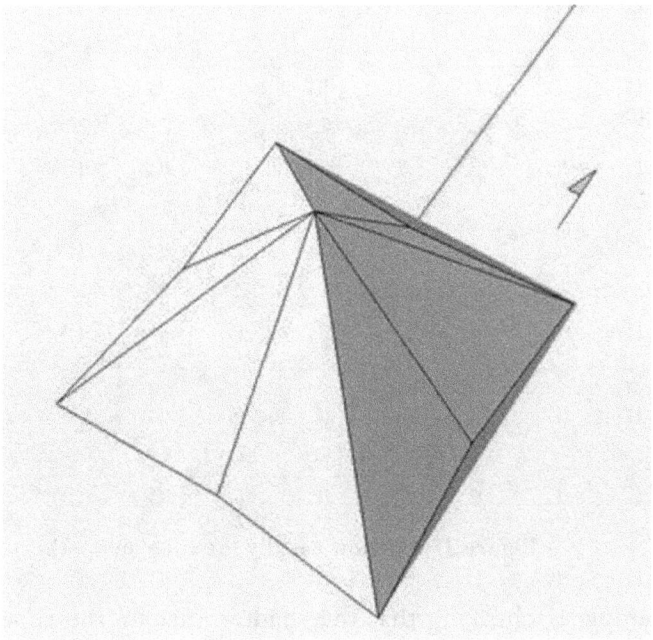

Figure 36: Concavity of the faces

The model also has its edges and apothems. When the shadow of one edge is cast on another on the model, the same thing must happen along the entire edge and on the base of the pyramid. The model is a reference, however, and its resolution is not enough for the perfection that was achieved. That required making the following control checks.

Control Checks

The procedure that we are describing allows us to make sure that the edges and the apothem meet at the summit point. This happens on the north face as well as on the rest of the faces of the pyramid. We can also ensure the straightness of the edges when they cast their shadows on the apothems. As regards the precision achieved in the plotting, it is possible to obtain it by making a number of checks.

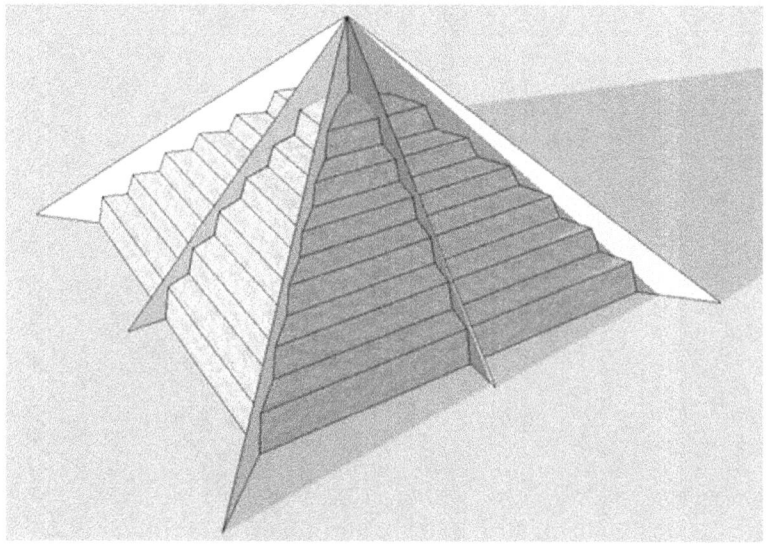

Figure 37: Shadow cast by the edge on apothem.

One example is checking that the shadow cast by the edges enters exactly by the N-W corner and leaves by the N-E corner. This guarantees that the corners of the base intersect the curved shadow path. Resolution at the base of the pyramid is 100 times greater than

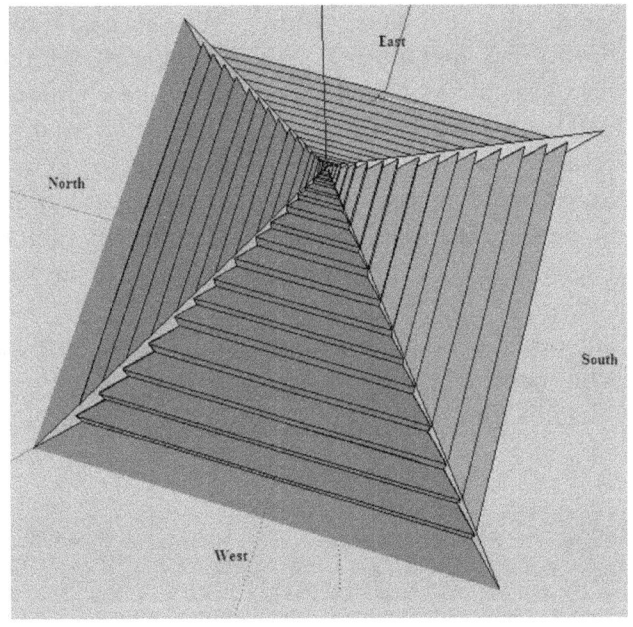

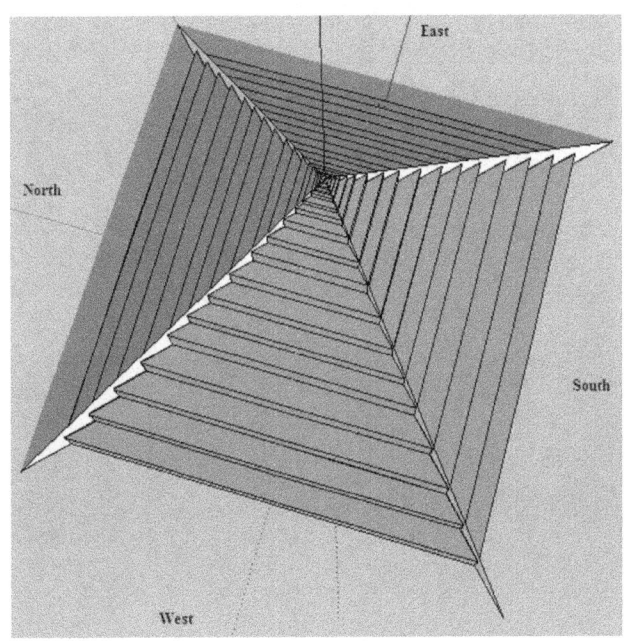

Figure 38: Control Checks.

that of the model. However, we must bear in mind that the procedure does not involve measuring but seeking perfection by spotting flaws. The sharpness of the shadow cast on these large screens (very long straight line) differs from the sharpness of the shadow cast by a gnomon (a point). We can say that while the gnomon casts the shadow of the summit point, the edges cast an infinite number of reference points. In addition, the use of horizontal screens on some steps make it possible to notice any plotting errors in even more detail. This procedure allows us to see any flaw in the plotting as if the casing had already been laid. These flaws are the kind that any observer could notice from the base of the pyramid. As we said at the beginning, the procedure they used is only useful for plotting pyramids, and the larger the pyramid, the easier it will be to spot errors.

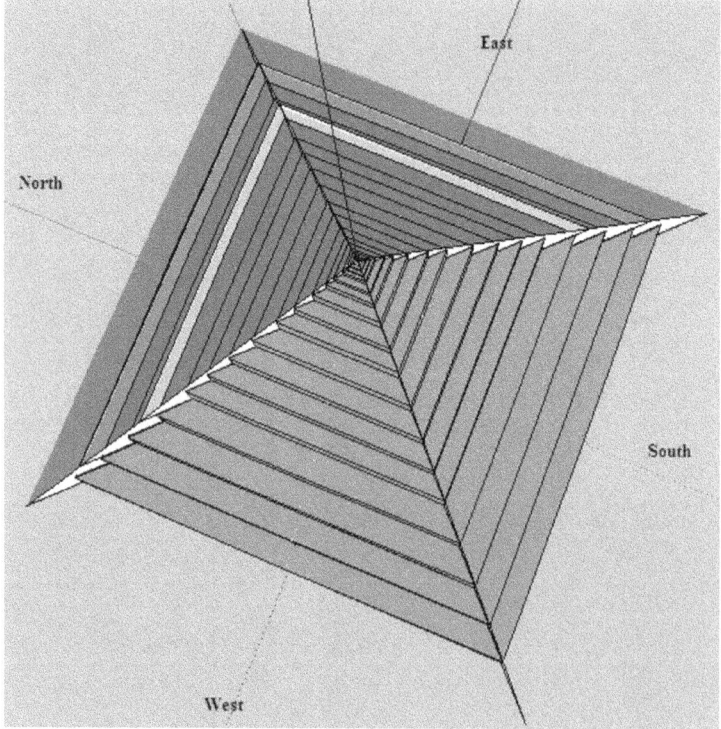

Figure 39: Horizontals Screens

Plotting the Rest of the Casing

The shadows cast by the pyramid throughout the year fall within the shadow area between the summer and winter solstices. These shadows make it possible to plot the north side of the pyramid, its edges and its apothem, as well as the apothems of the west and east sides. The edges and apothem on the south side fall outside the shadow area, and it will therefore be impossible to plot them by using the shadow cast directly by the model because it will never cast any shadows there (see Figure 29).

In the course of the year, the edges of the south side of the model cast their shadows on the edges of the north side. These shadows are equivalent to the shadows that will be cast by the edges of the south side of the pyramid on the edges of the north side of the pyramid and were used for plotting them. In the morning hours of October 11th, when the shadow touches the N-W corner, the shadow of the edges will mark out the north and west sides simultaneously. In the afternoon, at the moment the shadow leaves the pyramid by the N-E corner, the shadow of the edges will mark out the north and east sides.

The shadows cast on each other and on the apothems by the edges of the south side of the model are also equivalent to the same shadows of the pyramid and they were used for plotting the edges of the south side. The sun's declination is also present and it is reflected in the plotting of the east side of the pyramid of Khufu. The measurements that have been made of the Great Pyramid confirm those made by Flinders Petrie, with some small differences. According to Cole's measurements, the base is rotated counter-clockwise by an average 3'6 ".

The plotting of the base is not symmetrical in relation to the north-south axis. The west sector – plotted in the morning sun – is much more accurate; the angles of the corners are almost perfect: South-West 90° 0'33" + 0'33" and North-West 89° 59'58" -0'20" *.

The east sector – plotted in the afternoon hours – is less accurate due to the sun's declination; the angles of the corners have some difference with the right angle: North-East 90° 3 '20" +3 '20", South-East 89° 56'67" -3'33". This difference is between 3 '20" and -3'33", which is a result of the sun's declination.

Rotation has been exaggerated in the figure to make it more noticeable.

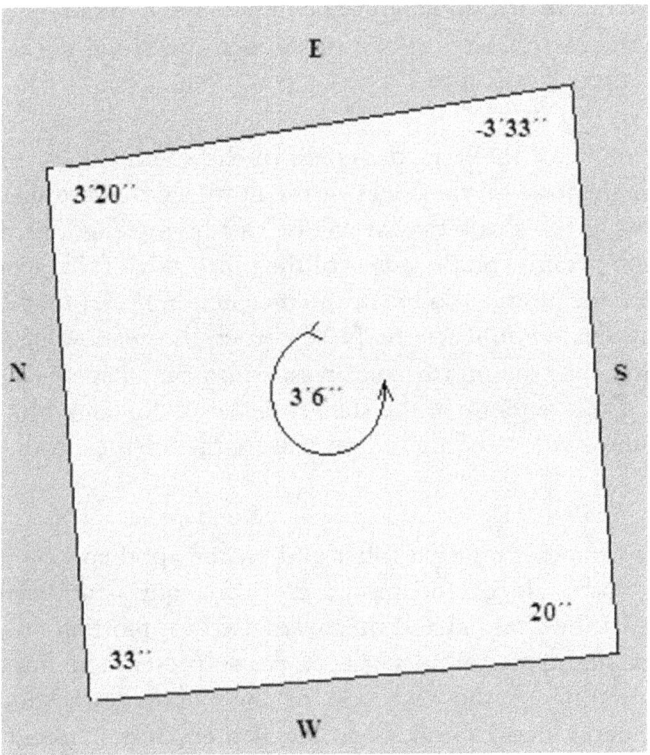

Figure 40: Measurements made by Cole

Laying the Casing

The Great Pyramid was built in the same way as the other pyramids: a stepped core was built first and then a casing was laid after the pyramidal shape had been plotted. The only new problem that arises is laying the casing at the top of the pyramid. The size of the casing stones decreases with height. The largest stones, which can weigh as much as a dozen tonnes, are found at the bottom, whereas those at the top do not exceed one tonne. This decrease in the size of the blocks is evidence of the difficulty of lifting them and the different techniques used to do so.

Ramps, which are simple embankments for each face, are used at the bottom. Beyond the height of the bottom levels, the main ramp was continued by connecting it with the south face. Once the stones had been manoeuvred up along the main ramp to the level at which the casing was being laid, they were moved on the surface of the casing under construction. The casing stones were raised beyond the average height of the pyramid by means of ropes, hoisting each block on the face of the pyramid. The blocks were accommodated on a wooden sledge that slid on wooden tracks to reduce friction. The effort was made by crews positioned on the steps of the core and on the surface of the casing under construction.

The Egyptian archaeologist Selim Hassan made a momentous discovery on the Giza plateau. He found a fixed bearing stone for ropes which has three parallel grooves in a half-round shape through which the strings slid. In the sector opposite the grooves there is a tenon-shaped projection with two holes whose purpose was to fasten the element to a structure by means of plugs (Verner 2001: 85).

A similar element was found in the pyramid of Khentkaus. It differs from the other one in that it has only one fastening hole in the tenon. This form of support makes it possible to modify the direction of the rope by at least 45 degrees.

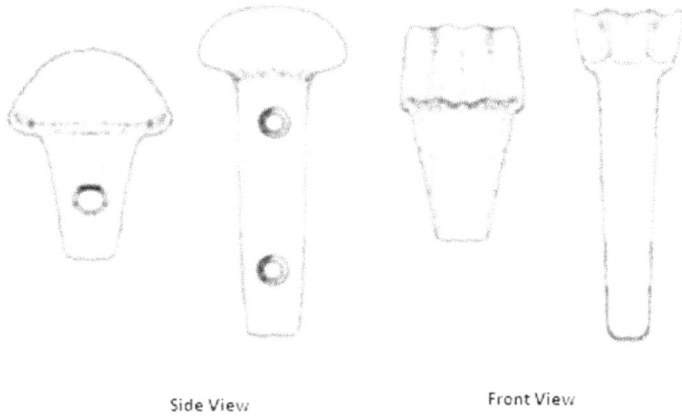

Side View   Front View

Figure 41: Bearing stones for ropes

Dieter Arnold shares the view that these bearing stones for ropes were attached to a wooden structure and adds that they "were part of an unknown device to pull or lower three parallel running ropes over an edge or around a corner" (Arnold 1991: 282).

Apart from the bearing stones for ropes, the use of counterweights assisted with the effort to be made. A counterweight is basically a container on a sledge that can be loaded with sand bags, chunks of stones and even the weight of the men themselves. The downward movement of the counterweight raises or helps raise the block, depending on the load of the counterweight, the weight of the block to be raised and the losses from the rope rubbing on the bearing stones.

The Purpose of the Grand Gallery

The question that arises here is whether the counterweight used was inside or outside the pyramid. The reason for this question is that there is a large ramp inside the Great Pyramid called the Grand Gallery. The chambers and corridors that make up the interior of the pyramids were initially distributed underground and then incorporated into the structure at its bottom.

The design of the Great Pyramid is atypical in that the chambers and corridors grew in height inside the structure. The reason for this unusual design is that it included the Grand Gallery. The Grand Gallery is an indoor ramp or inclined plane in the shape of a gallery because it is inside the structure of the pyramid core. This gallery has the necessary infrastructure for a counterweight to be slid along inside it. It has two stone tracks at floor level next to the walls. Another interesting detail is the stone blocks embedded in the walls at regular intervals and capable of stopping the counterweight in intermediate positions. (4)

The Grand Gallery acted as a corridor leading to the upper chamber. However, its design has characteristics that cannot be accounted for by this purpose alone.

"There are unique features within this gallery that deserve comment, features which for centuries have perplexed researchers." It is necessary to have "a clear understanding of all of the pieces of the puzzles ... to explain the purpose of the gallery in its relationship to the pyramid as a whole." To date, no pyramid scholar has been able to explain the existence of the Grand Gallery or its peculiarities. (Lepre 1990: 79).

Location on the north-south plane: The Grand Gallery has the peculiarity of ending exactly on the central plane of the pyramid. In order for the Grand Gallery to serve the purpose of containing a counterweight, the rope that transferred the effort towards the outside had to rise vertically from the top of the Grand Gallery to reach the summit through a shaft that we will call 'rope shaft.'

Location on the east-west plane: The Grand Gallery was built on a plane displaced from the north-south central plane by 7.5 meters to the east. Asymmetry is an unusual element in Egyptian architecture, which used symmetry as a prevailing element. This asymmetry in the location of the Grand Gallery is for a significant reason in the design which confirms its purpose. (5)

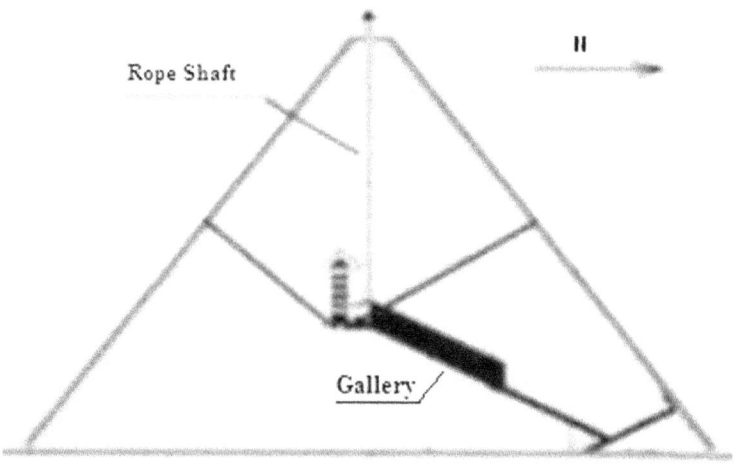

Figure 42: Location of the Gallery on the north-south plane

The existence of very long shafts of small section running across the masonry of the core is a characteristic of this pyramid. If we project the upper sector of the Grand Gallery – where the big step is located – vertically towards the top of the pyramid, we can see that the rope shaft that is needed to transfer the effort to the outside would come out at the side of the pyramidion.

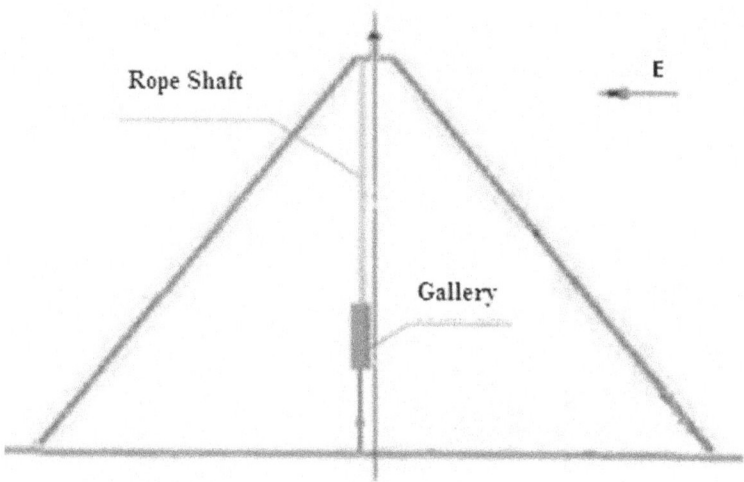

Figure 43: Location of the Gallery on the east-west plane

The location of the Grand Gallery displaced from the central plane is evidence of the main purpose for which it was included in the design, namely to contain a counterweight used during the laying of the casing in the middle and upper sectors of the pyramid. The figure is a drawing of the summit of the pyramid by E.W. Lane, a professional draughtsman, in his work "Exhaustive Description of Egypt" (British Museum, add. MS. 34083, f. 24), published by C.W. Ceram in his book "In search of the past."

The figure shows a rectangular sector with characteristics similar to those of the mouth of the shaft, in the aforementioned location, containing three small stone blocks (see arrow).

The following website makes it possible to visit the top platform and confirm the accuracy achieved by Lane in his drawing   (see: http://www.pbs.org/wgbh/nova/ancient/explore-ancient-egypt.html).

Once the casing was in place, the outer layer came from the quarry with an uneven shape that was given a finish on site. Horizontal strings were laid from the edges on the area where the casing was being finished.

Figure 44: The top of the Khufu´s pyramid

The measure of a strip with markings that indicated the distance from the horizontal string to the surface of the casing was taken as a reference.

The vault construction of the Grand Gallery (as can be seen in the chambers and corridors of the pyramid at Meidum) ends where the walls meet, above the existing horizontal roof. In January 2017 I published an article on Academia.edu (The moderns research about the empty spaces inside the Great Pyramid) explaining my opinion about the existence of a large hollow space above the roof of the Grand Gallery.

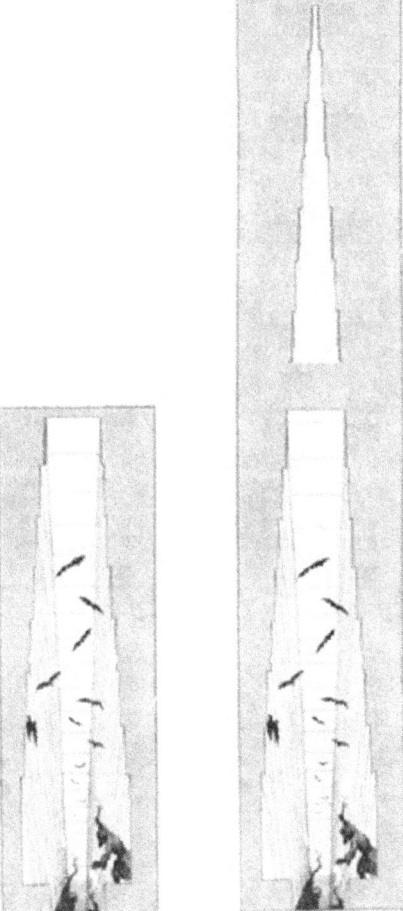

Figure 45: The Grand Gallery

The Khafre Pyramid was built right after Cheops´s Pyramid and is slightly lower in height; it is logical to deduce that its inner layout should be similar to the one we have just analyzed.

Bibliography

Dieter Arnold, Building in Egypt: Pharaonic Stone Masonry, Oxford University Press, 1991.

Borchardt Ludwig. Das Grabdenkmal des Koniges Sahure, vol I. Leipzig: J. C. Hinrichs, 1910.

Ceram C. W., En Busca del Pasado, Labor , 1961.

Colonel Coutelle, Observaciones sobre las Pirámides de Gizeh, vol. IX Description de l'Égypte, Paris 1829.

Dash, Glen, "North by Northwest: The Strange Case of Giza's Misalignments," AERAGRAM, Vol. 13 no. 1 (Spring 2012), 10-15

Dash, Glen, "New Angles on the Great Pyramid," AERAGRAM, Vol. 13 no. 2 (Fall 2012), 10-19.

(1) Dash, Glen , Did the Egyptians Use the Sun to Align the Pyramids?

(2) Dash, Glen, "Simultaneous Transit and Pyramid Alignments: Were the Egyptians' Errors in Their Stars or in Themselves? AERAGRAM (January 27, 2015).

Dash, Glen, "How the Pyramid Builders May Have Found Their True North," AERAGRAM, Vol. 14 no. 1 (Spring 2013), 8-14.

Edwards I.E.S., The Pyramids of Egypt, Penguin Books, 1993.

Fakhry Ahmed, The Pyramids, The University of Chicago Press, 1975.

(4) Gerardo Daniel, Construction at Giza, Magazine of Uruguayan Insitute of Egyptology, (1/1981).

(5) Gerardo Daniel, La Pirámide Posible, Amazon, (11/2012).

Hawass Zahi, Pyramid Construction,New Evidence Discovered at Giza, http://guardians.net/hawass/pbuildrs.htm"

Hawass Zahi and Mark Lehner, The Pyramids of Giza, 2017.

Isler Martin, Sticks, Stones, and Shadows: Building the Egyptian Pyramids, 1926

Lauer J.P, Le Problème des Pyramides D`Égypte, Payot, 1948.

Lehner Mark, The Complete Pyramids, Thames & Hudson, 1997.

Lepre J.P., The Egyptian Pyramids, Mc Farland, 1990

Maragioglio Vito and Celeste Rinaldi. L'Architettura delle Piramidi Menfite, Rapallo, 1965.

Mendelssohn Kurt, The Riddle of the Pyramids, Thames and Hudson, 1974.

Petrie Flinders, The Pyramids and Temples of Gizeh, Scribner & Welford, 1883.

Sampsell Bonnie M., Pyramid Design and Construction - Part I: The Accretion Theory, The Ostracon, Journal of the Egyptian Study Society, Denver, 2000.

SmythCraig B., How the Great Pyramid was built, Smithsonian Books, 2006.

Miroslav Verner, Las Pirámides, El Misterio, Cultura y Ciencia de los grandes monumentos de Egipto, Grove Press, 2001.

ABOUT THE AUTHOR

Daniel Gerardo was born in Montevideo on the 25th of November 1958. He is a Mechanical Engineering Expert.

From a young age, he has had a particular inclination for the study of the evolution in the construction of the Egyptian pyramids. It was in a history lesson at the age of 12 that he saw a sectional drawing of Pharaoh Khufu's pyramid for the first time.

In the 1980s he completed his thesis, which dealt with the relationship between the design of the interior distribution of the Great Pyramid and its construction. The work was assessed by architect J. F. Lauer. In 1981 he published the article "Construction at Giza" in the journal of the Uruguayan Institute of Egyptology.

He continued to work on his research based on the analysis of the functional and construction requirements related to the creation of this masterpiece and the available procedures to meet them.

In 2012 he published his book "The Feasible Pyramid," which discusses the plotting of the great pyramids. In 2013 he was awarded the Medal of Merit of the Arab Republic of Egypt in a contest organized by the Egyptian Embassy and the Uruguayan Institute of Egyptology.

The conclusions of this extensive research were published in this book "Plotting and Building the Pyramids of Giza", together with the opinions of leading specialists on the subject.

DANIEL GERARDO

www.ingramcontent.com/pod-product-compliance
Lightning Source LLC
Chambersburg PA
CBHW071618170526
45166CB00003B/1101